COLLECTING
Yo-Yos

COLLECTING
Yo-Yos

Prof. James L. Dundas

4880 Lower Valley Road, Atglen, PA 19310 USA

Copyright © 2000 by James L. Dundas
Library of Congress Catalog Card Number: 99-64844

All rights reserved. No part of this work may be reproduced or used in any form or by any means—graphic, electronic, or mechanical, including photocopying or information storage and retrieval systems—without written permission from the copyright holder.

"Schiffer," "Schiffer Publishing Ltd. & Design," and the "Design of pen and ink well" are registered trademarks of Schiffer Publishing Ltd.

Design by Blair R. Loughrey
Type set in ZurichUblkexbt / Zurich BT

ISBN: 0-7643-1010-0
Printed in China
1 2 3 4

The yo-yos, advertisements, characters, symbols, and logos presented in this book are all trademarked and/or copyrighted properties. None of the copyright/trademark holders in any way authorized this book, nor furnished or approved any of the information contained therein. This book is for the information, use, and entertainment of its readership. This book in no way attempts to infringe on any copyright or trademark.

Published by Schiffer Publishing Ltd.
4880 Lower Valley Road
Atglen, PA 19310
Phone: (610) 593-1777; Fax: (610) 593-2002
E-mail: Schifferbk@aol.com
Please visit our web site catalog at
www.schifferbooks.com
or write for a free catalog.
This book may be purchased from the publisher.
Please include $3.95 for shipping.

In Europe, Schiffer books are distributed by
Bushwood Books
6 Marksbury Rd.
Kew Gardens
Surrey TW9 4JF England
Phone: 44 (0)181 392-8585; Fax: 44 (0)181 392-9876
E-mail: Bushwd@aol.com

Please try your bookstore first.

We are interested in hearing from authors with book ideas on related subjects.

DEDICATION

This book is dedicated to my brother, and all of those brothers and sisters who were killed in Vietnam.

BROTHER

Only memories of you my brother have I left to ease the pain,
only memories as of sunshine that is banished by the rain.
With the autumn flowers you faded and with them have gone to rest.
Forgive me in my anguish, for perhaps thou knowest best.
All in vain I try to fathom why you went to Vietnam,
was your gallant soul concealing, what we know so well today?
That you were a true American and believe in the American way,
and now have gone away.
Winter's winds will sigh in mourning, spring will bring the flowers once more,
to embellish summer's grandeur, then to die at autumn's door.
Some day God will call to me, and I will set forth,
but perhaps it's a sunny morning and I will find a garden fair,
full of flowers and little children, and my Brother will be there.
Smiling he will come to meet me, and I think I hear him say.
"Did you say that you had missed me? Why I haven't been away."

For Sgt. Jerry R. Dundas
Killed at Khesanh, Vietnam on 5/5/68
from his brother Prof. James L. Dundas

And to my dear wife Janet M. Dundas who started my interest in old toys.

CONTENTS

Acknowledgments	8
Introduction	9
History	10
Patents	12
All the Yo-Yos	14
All Western Plastic	14
Alox	15
Bandalor	16
Big Chief	16
Bob Allen	16
Champion	17
Cheerio	17
Chemtoy	20
Chico	21
Conestoga Corp.	21
Cracker Jack	22
Dell	22
Duncan	23
Festival	72
Fli-Back	75
Flores	81
Forrester	81
German	82
Goody	82
Gropper	86
Hasbro	87
Hi-Ker	88
Hy-Lo	91
Jack Russell	91
Ka-Yo	92
Kaysons Novelty Co.	93
Knights	94
Kusan	94
L.H. Knibbco	95
Lumar	95
Mary	96
Medalist	96
Miscellaneous	97
N.N. Hill Brass co.	107
Nadson	107
Novel Swing Top	108
Parker	108
Royal	109
Star	111
Strings	111
Tico	123
Whirl King	124
World's Fair	125

ACKNOWLEDGMENT

A very special thanks to Jason Colwell and Julie Martensen for their help and the use of some of their yo-yo's.

INTRODUCTION

In recent years, interest in yo-yo collecting has grown rapidly, as has the value of collectible yo-yos. Yo-yos that sold for 50 cents in the 1940s and 1950s can be worth hundreds of dollars, and the prices are only going up. There has long been an interest in antique toys from the 1800s to the 1930s, particularly in cast iron toys like banks. But now they cost from hundreds to thousands of dollars. The toys of the 1940s through the 1960s are eagerly sought after today not only for nostalgic reasons but for the money that can be made in selling them.

Unfortunately most of us were not able to keep everything from our childhood. Those of us born after 1945, the "baby boomers," are now looking for the toys that we grew up with. I want the toys that my mother threw out or I lost (or broke) when I was a kid. If you have a display of toy yo-yos that you can show your friends, you will frequently hear, "I had one of those."

I remember my first yo-yo — it was the Spring of 1950 and I was a safety patrol boy stationed on the corner next to Cammas's Candy Store. A man approached me, asking if I would tell the kids about a yo-yo contest after school at 3:30 today. I said I would and he asked, "Are you sure?" I said, "Yes, I will." At that point he opened a box; it was full of brand new yo-yos. They were Duncan four jeweled yo-yos. I couldn't decide between the red or dark blue model. I took the blue one. At 3:30 I was there on the corner with at least 30 other kids. The contest began. I didn't even bother to enter because I didn't know how to work a yo-yo! Though I didn't compete, I still had a free yo-yo.

Yo-yos are a low-keyed collectible. You can still find some with low prices, although others have increased a great deal. I think that after reading this book you will find a lot of pleasure and reward in collecting different types of yo-yos. There are hundreds of them out there!

Professor James L. Dundas

HISTORY

The toy yo-yo has been enjoyed for centuries in the Philippines as well as various countries around the world. Commonly know as a Bandalore, it was introduced in the United States in the mid-1860s by James Haven and Charles Hettrick. However, the term yo-yo was first used in the U.S. in 1928 by Pedro Flores, an immigrant from the Philippines. Flores built the first yo-yo factory in Santa Barbara, California, and began running yo-yo contests that year. Flores sold the yo-yo trademark to Duncan in 1930; Duncan's yo-yo contests and promotions soon made the Duncan yo-yo famous across the U.S.

Some of the other major yo-yo companies and brands included:

Cheerio:

Cheerio yo-yos were made by the Kitchener Buttons Limited Factory in Kitchener, Ontario, Canada, by Wilfred Schlee, Sr. The firm first produced yo-yos in 1931. While the firm's first yo-yos were called Hi-Ker, they were soon overtaken by a new model named Cheerio. Cheerio entered the U.S. market in 1946 and became a major competitor for Duncan. Cheerio sold its American operation to Duncan in 1954; Duncan continued to produce a Cheerio line until the early 1960s. The Canadian production also ended in the early 1960s.

Duncan:
The company began in 1929 in Chicago as Donald F. Duncan, Inc. Duncan's lawsuit against Royal over the "yo-yo" trademark resulted in Duncan losing its exclusive rights to the term "yo-yo" in 1965. Duncan declared bankruptcy that same year and later sold out to Flambeau Plastics, the firm that continues to produce Duncan yo-yos today.

Festival:
The Festival Yo-Yo Company began in 1965 and was bought out by the Union Wadding Company in 1966. Monogram Products Inc. acquired the Festival line in 1981 and continues production today.

Fli-Back:
The Fli-Back Company, founded by James Gibson in the early 1930s, began making yo-yos in 1946. The Ohio Art Company bought Fli-Back in 1968 and continued manufacturing Fli-Back yo-yos through the 1970s.

Goody:
The Goody Manufacturing Company in New York began making "Filipino Twirlers" prior to World War II and continued to produce them well into the 1960s.

Royal:
The Royal Tops Manufacturing Company was started in 1937 by Joe Radovan and continued production into the early 1980s.

PATENT

Patent Numbers can be a great help in determining the age of an object. If a patent number is 2,185,170 to 2,227,418 then the object can be no older than 1940. Keep in mind that the object may still be made today, however. But this is usually not the case; patent numbers expire after 17 years. Whether or not an item has an area code is another way to tell the approximate age of an object. Mr. Walker, the postmaster in New York, introduced a two number area code for large cities on May 6, 1943, and stopped using the two digit codes when zip codes were added to the mail. Zip codes, using five numbers, started on July 1, 1963.

If an object with an address has no area or zip code, then the item may be older than May 6, 1943. However, in a town or city that had only one post office from 1943 to 1963, there were no two number area or zip codes. In October 1984, Zip + four came into being; it has the five number Zip code plus four more numbers.

So, if you have something with a two number area code, then it was made from 1943 to 1963. If it has a zip code then it was made from 1963 to 1984. If you have something with five numbers then four numbers, it can be no older than October 1984.

When you find the words "Made in Japan" printed on something, then it was made from 1921 to 1942 and maybe to the end of World War II, Aug. 1945. If you find "Occupied Japan" printed on something, then it was made from 1945 to 1952. Japan was occupied by the U.S. Army at the end of World War II, from late August 1945 to April 28, 1952.

Here is an example:

John Doe Toys
123 Main Street
Detroit Mich. (It may be older than May 6, 1943.)

John Doe Toys
123 Main Street
Detroit 09 Mich. (It was made from 1943 to 1963.)

John Doe Toys
123 Main Street
Detroit 48207 Mi. (It was made from 1963 to 1984.)

John Doe Toys
123 Main Street
Detroit 48207 -
1144 MI (It was made from 1984 to Date.)

PATENT NUMBERS

DATE	Patent #									
		1866	51784	1898	596467	1930	1742181	1962	3015103
		1867	60658	1899	616871	1931	1787424	1963	3070801
1836	1	1868	72959	1900	640167	1932	1839190	1964	3116487
1837	110	1869	85503	1901	664827	1933	1892663	1965	3163365
1838	546	1870	98460	1902	690385	1934	1941449	1966	3226729
1839	1061	1871	110617	1903	717521	1935	1985678	1967	3295143
1840	1465	1872	122304	1904	758567	1936	2026516	1968	3360800
1841	1923	1873	134504	1905	778834	1937	2066309	1969	3419907
1842	2413	1874	146120	1906	808618	1938	2104004	1970	3487470
1843	2901	1875	158350	1907	839799	1939	2142080	1971	3551909
1844	3395	1876	171641	1908	875679	1940	2185170	1972	3633214
1845	3873	1877	185813	1909	908430	1941	2227418	1973	3707729
1846	4348	1878	198753	1910	945010	1942	2268540	1974	3781914
1847	4914	1879	211078	1911	980178	1943	2307007	1975	3858241
1848	5409	1880	223211	1912	1013095	1944	2338061	1976	3930271
1849	5993	1881	236137	1913	1049326	1945	2366154	1977	4000520
1850	6961	1882	251685	1914	1083267	1946	2391856	1978	4065812
1851	7865	1883	269320	1915	1125212	1947	2413675	1979	4131952
1852	8622	1884	291016	1916	1166419	1948	2433824	1980	4180876
1853	9512	1885	310163	1917	1210389	1949	2457797	1981	4242757
1854	10358	1886	333494	1918	1251458	1950	2492944	1982	4308622
1855	12117	1887	355291	1919	1290027	1951	2536016	1983	4366579
1856	14009	1888	375720	1920	1326899	1952	2580379	1984	4423523
1857	16324	1889	395305	1921	1354054	1953	2624046	1985	4490885
1858	19010	1890	418655	1922	1401948	1954	2664562	1986	4562596
1859	22477	1891	443987	1923	1440352	1955	2698434	1987	4633526
1860	26642	1892	466315	1924	1478996	1956	2728913	1988	4716594
1861	31005	1893	483976	1925	1521590	1957	2775762	1989	4794652
1862	34045	1894	511744	1926	1568040	1958	2818567	1990	4890335
1863	37266	1895	531619	1927	1612700	1959	2868973	1991	4980927
1864	41047	1896	552502	1928	1654521	1960	2919443	1992	5077836
1865	45685	1897	574369	1929	1696897	1961	2966681		

ALL THE YO-YOs

All Western Plastics

Roy Rogers yo-yo, assorted colors with extra strings, made of plastic by All Western Plastics in the 1950s. $15-25

Alox Flying Disc, gold die stamp, some two-tone, some one-color, made of wood in the 1950s-1960s. $15-20

Alox

Alox Flying Disc Tournament model, yellow paint seal, made of wood in the 1950s-1960s. $20-30

Alox Flying Disc, navy-green, made of wood in the 1950s-1960s. $15-25

Bandalor Co.

Bandalor Co. paper seal adorned yo-yo. This yo-yo was produced by Pedro Flores, known for being the first person to introduce yo-yos in America. Flores started the Bandalor Company after he left Duncan. This example was made of wood in the 1930s-1940s. $250-300

Big Chief

Big Chief Tournament, swirled, made of plastic in the 1950s. $20-30

Bob Allen

Bob Allen Sidewinder, red hot stamp on white sided black yo-yo, on card, made of plastic in the 1960s. Wood and tin models were also produced. $40-50 (plastic)

Cheerio

Cheerio 55 Beginners, silver foil seal, two-tone yo-yo made of wood in the 1950s-1960s. $75-85

Champion

Champion Style 55, gold die stamp, made of wood in the 1960s. $15-20

Cheerio Champion, silver disc with eagle symbol on each side, made of wood in the 1950s. $200-250

Cheerio Glitter Spin, silver foil seal, 4 jewels on each side. Several seal versions were made. This wooden, jeweled yo-yo was made in the 1950s. $200-250

Cheerio 55 Beginners, die stamp, two-tone, made of wood in the 1950s. $65-75

Cheerio Rainbow Pro, model die stamped, airbrush stripe, made of wood in the 1950s. $125-150

Cheerio Pro 99, gold foil seal, natural, made of wood in the 1950s. $85-95

Cheerio Pro 99, gold foil seal, metallic paint (also came with gold flake paint and matte paint), made of wood in the 1950s. $85-95

CHEERIO - CHEMTOY

Cheerio Pro 99, gold foil seal, blue with black strip paint, made of wood in the 1950s. $85-95

Cheerio Big Chief, matte paint, made of wood in the 1950s. $85-95

Chemtoy

Chemtoy yo-yo made of tin in Hong Kong in the 1950s. $30-40

Chico

Chico Super Tournament Top, gold paper seal, made of wood in the 1950s and 1960s. $60-80

Conestoga Corp.

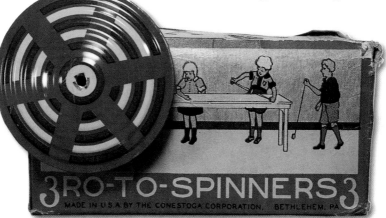

Conestoga Corp. "3 Ro-To-Spinners" (yo-yo called Ro-To-Bob) box and top, zing-zong & yo-yo, all tin litho in box, made in the 1930s. $200-250

Cracker Jack Co.

Above: The Cracker Jack Co. teeny has the Cracker Jack name on one side and Sailor Jack on other. The teeny was made of plastic in the 1960s. $20-30

Right: Cracker Jack mini prize yo-yo, red and black, made of plastic in the 1960s. $15-20

Dell

Above: Dell Big "D" Flying Star butterfly model, red yo-yo with black sides, made of plastic in the 1960s. $30-40

Right: Dell Big "D" Trickster, swirled plastic with hot stamp, carded, made of plastic in the 1960s. $25-35

Duncan

Genuine Duncan No. 88 Whistling Yo-Yo propeller whistler, second series. This whistler yo-yo was made of tin in the 1930s. $150-175

Genuine Duncan No. 88 Whistling Yo-Yo, red and black pinwheel whistler, first series. This whistler yo-yo was made of tin in the 1930s. $175-200

Genuine Duncan No. 88 Whistling Yo-Yo Hypno Swirl whistler, first series. This whistler yo-yo was made of tin in the 1930s. $175-200

Genuine Duncan No. 88 Whistling Yo-Yo Dart Board whistler, second series, made of tin in the 1930s. $145-165

Genuine Duncan No. 88 Whistling Yo-Yo Ink Blot whistler, first series, made of tin in the 1930s. $175-200

Genuine Duncan No. 88 Whistling Yo-Yo Starburst whistler, second series, made of tin in the 1930s. $145-165

Left: Genuine Duncan No. 88 Whistling Yo-Yo, solid green Big G whistler. "Big G" refers to the "*G*" in the word "Genuine" wrapping around the entire seal. This whistler yo-yo was made of tin in the 1940s. $100-150

Center: Genuine Duncan No. 88 Whistling Yo-Yo, solid blue Big G whistler, made of tin in the 1940s. $100-150

Right: Genuine Duncan No. 88 Whistling Yo-Yo, solid orange Big G whistler, made of tin in the 1940s. $100-150

Genuine Duncan No. 22 Rainbo Yo-Yo, color changer inner spinning disc, made of tin in 1934 (patented). $300-350

O-Boy Duncan Whistling Yo-Yo, litho, jr. size, black, at least 3 color combos and 2 seal variations, made of tin in the 1930s. $125-150

O-Boy Duncan Whistling Yo-Yo, litho, jr. size, green, at least 3 color combos and 2 seal variations, made of tin in the 1930s. $125-150

Duncan, Little G Genuine Tournament, small *G* die stamp, made of wood in the 1940s-1950s. $45-60

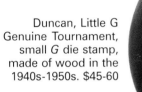

Duncan, Big G Genuine Duncan Yo-Yo, big *G* die stamp. It doesn't say "Tournament" on this wooden yo-yo made in the 1930s. $65-75

Duncan, Big G Genuine Tournament, big *G* die stamp, made of wood in the 1930s. $50-65

Duncan, Big G Yellow Seal Genuine Tournament seal, big *G*, yellow, made of wood in the 1930s. $70-80

Duncan, Big G Gold Seal Genuine Tournament, gold seal, big *G*, gold, made of wood in the 1930s. $70-80

DUNCAN

Duncan's, O-Boy wood, larger *D* and *N* in Duncan, no Pat. Pend, die stamp, made of wood in the 1930s. $45-55

Above: Duncan's, O-Boy Yo-Yo Patent Pending, wood, made in the 1930s. $65-75

Right: Genuine Duncan Junior No. 33 Yo-Yo Tops, red and black, big *G*, undersized, made of wood in 1930s-1940s. $20-30

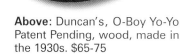

Genuine Duncan Beginners No. 44 Yo-Yo Tops, red and black, small G, full sized, made of wood in the 1950s-1960s. $15-20

Above: Genuine Duncan Beginners No. 44 Yo-Yo Tops, red and black, big G, made of wood in the 1930s-1940s. $20-30

Right: Genuine Duncan Beginners No. 44 Yo-Yo Tops, red and black, small G, undersized, made of wood in the 1950s-1960s. $15-20

DUNCAN

Duncan Special 44 Yo-Yo Top, two-tone (blue and red), made of wood in the 1950s. $25-35

Duncan Beginners, blue and red, Reg. U.S. Pat. Off., made of wood in the 1960s. $35-45

Duncan Junior Yo-Yo, undersized, italic Junior, two-tone, die stamp, made of wood in the 1960s. $20-30

Duncan Beginners Yo-Yo, yo-yo man's face with stars for eyes, word "Beginners" for mouth. This is one of the more rare Duncan Beginners yo-yos, made of wood in the 1960s. $35-45

Duncan's Junior, word "Junior" italicized, red and blue, made of wood in the 1960s. $20-30

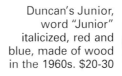

Duncan Beginners, pink, Reg. U.S. Pat. Off., made of wood in the 1960s. $35-45

Duncan Junior Yo-Yo, red and yellow, made of plastic in the 1960s. $20-30

DUNCAN

Genuine Electric Lighted Duncan Yo-Yo Tops, plastic, lights up with batteries, big *G*. This is the first battery-operated Duncan light-up yo-yo made of plastic in the 1950s. $65-75

Genuine Duncan No. 77 Yo-Yo Tournament Top, yo-yo man on the yellow seal, made of wood in the 1950s. $65-75

Genuine Duncan Tournament Yo-Yo Tops, yellow seal with small *G*, made of wood in the 1950s. $65-75

Duncan Tops Yo-Yo, silver slanted writing, two-tone yo-yo made of wood in the 1930s. $50-75

Duncan Super Yo-Yo Tournament Tops, gold die stamp. In 1996, a reproduction was issued of this model. You can tell the reproduction from the original by looking at the registration mark. The reproduction has the (R) next to the word Duncan, while the original has the (R) below the hyphen in the word yo-yo. This original yo-yo was made of wood in the 1950s. $25-30

Duncan Super Yo-Yo Tournament Tops was made of wood in the 1950s. $30-40

Duncan Super Yo-Yo Tournament Return Tops. This was a later model than the version that does not say "return." This model was made of wood in the 1950s. $40-50

Duncan Super Tournament Tops, jeweled model, metallic paint, says "super" but not "jeweled," made of wood and jewels in the 1950s. $75- 85

Duncan, Genuine Tournament Yo-Yo Duncan Tops, jeweled model (does not say jeweled), metallic paint, made of wood and jewels in the 1950s. $90-110

Duncan Jeweled Tournament Yo-Yo Tops, jeweled model, metallic paint, made of wood and jewels in the 1950s-1960s. $65-75

Duncan Jeweled Tournament Yo-Yo Tops, jeweled model with dark blue matte paint, made of wood and jewels in the 1950s-1960s. $65-100

Duncan Jeweled Tournament Yo-Yo Tops, jeweled model with red matte paint, made of wood and jewels in the 1950s-1960s. $65-100

Duncan Jeweled Tournament Yo-Yo Tops, jeweled model with pink matte paint, made of wood and jewels in the 1950s-1960s. $65-100

Duncan Jeweled Tournament Yo-Yo Tops, light blue, made of wood and jewels in the 1950s-1960s. $65-100

Duncan Jeweled Tournament Yo-Yo Tops, plain wood, made of wood and jewels in the 1950s-1960s. $65-100

Duncan oversized tournament, 4 1/2" wide x 2" thick, made of wood in the 1950s. $150-200

Duncan No. 888 Pearlessence Tournament, metallic paint. Unlike many Duncan models, those with metallic paint do not feature an airbrushed stripe. This yo-yo was made of wood in the 1950s. $65-75

Luck-E JA-DO Contest Top, four-leaf clover design, gold die stamp. This is one of a few Duncan yo-yos that does not have the name "Duncan" on it. This yo-yo was made of wood in the 1950s. $60-70

Luck-E JA-DO Contest Top, dark wood. This is one of a few Duncan yo-yos that does not have the name "Duncan" on it. This yo-yo was made of wood in the 1950s. $60-70

Duncan Tenite Imperial Yo-Yo, translucent, also came without the word "Tenite" on it. The stamp is known as a chevron imperial stamp. Later versions of this yo-yo have a fleur de lis stamp. This model is still being made. A plastic yo-yo produced in the 1950s. $30-40

Duncan Tenite Imperial Yo-Yo, opaque (more rare than translucent), also came without word "Tenite." This stamp is known as a chevron imperial stamp; the later version of this yo-yo with the fleur de lis stamp is still being made. This 1950s yo-yo was made of plastic. $40-50

Duncan No. 22 Pony Boy Yo-Yo Top, undersized, BBs inside make the yo-yo rattle, clear edges, made of plastic in the 1950s. $50-60

Duncan No. 22 Pony Boy Yo-Yo Top, undersized, BBs inside make yo-yo rattle, green, made of plastic in the 1950s. $50-60

Duncan Butterfly Yo-Yo, butterfly shape, metal flake paint, gold die stamp with a picture of a butterfly, made of wood in the 1950s-1960s. $35-45

Duncan advertising yo-yo: the Duncan Kitty Clover Potato Chips Yo-Yo Return Top. This is an undersized beginner's model with writing on one side only. It was made of wood in the 1960s. $10-20

Duncan advertising yo-yo sold in Coast-to-Coast Stores. This is an undersized beginner's model with a beginners die stamp on back. It was made of wood in the 1960s. $10-20

Left: Duncan advertising yo-yo; Red Dot (potato chips), an undersized beginner's model featuring a beginners die stamp on back. It was made of wood in the 1960s. $15-25

Below: Duncan advertising yo-yo: Safe-T cones-cups, undersized beginner's model with a beginners die stamp on back. This model was made of wood in the 1960s. $15-25

Duncan advertising yo-yo: Bosco Bear Yo-Yo. This is an undersized beginner's model with a beginners die stamp on back. It was made of wood in the 1960s. $15-25

Duncan advertising yo-yo: So Good Potato Chip Co., undersized beginner's model made of wood in the 1960s. $15-25

Duncan advertising yo-yo: Schweigert Makes It Better, undersized butterfly beginner's model with a beginner die stamp on back. This yo-yo was made of wood in the 1960s. $15-25

Duncan advertising yo-yo: Stoppenbach Meats of Excellence since 1875, undersized butterfly beginner's model with a beginners die stamp on back. This yo-yo was made of wood in the 1960s. $15-25

"10 2 4 Dr. Pepper." Although this yo-yo doesn't say Duncan on it, it was made by Duncan for Dr. Pepper. Die stamped, this model was made of wood in the 1950s. $25-35

Duncan Genuine yo-yo, a mini yo-yo about 1 3/4" diameter, gold die stamp, came with a gift pack. This mini yo-yo was made of wood in the 1950s. $20-30

Duncan Tournament (jeweled cross-flags) on card, jeweled model made of wood in the 1960s. $50-85

Duncan cross flag Tournament, on card, originally 49 cents, orange yo-yo, made of wood in the 1960s. $50-85

Duncan cross flag Tournament on card, package marked 59 cents, green yo-yo, made of wood in the 1960s. $50-85

Duncan Tournament Yo-Yo, black flake paint with silver stamp, ornate lettering for tournament, made of plastic in the late 1960s. $20-30

Duncan Butterfly on a 69-cent card, black flake paint, made of wood in the 1960s. $35-60

Duncan Personalized butterfly pack with trick book, 2 strings (2 for 10 cents), and gold letters. This butterfly yo-yo was made of wood in the 1960s. $85-95

Duncan Beginners yo-yo features a picture of the yo-yo man's head on the toy. The yo-yo remains on the card and originally sold for 39 cents. The yo-yo is green-black and was made of wood in the 1960s. $10-15

DUNCAN

Duncan Beginners yo-yo features a picture of the yo-yo man's head. The yo-yo remains on the card and originally sold for 29 cents. This yo-yo is purple blue and was made of wood in the late 1950s-1960s. $10- $15

Duncan Satellite Mark 1, on the card, originally sold for 69 cents, white slanted writing, flat sides, made of wood in the 1960s. $60-70

Duncan Satellite Mark 1, round sides, pink, planets and stars on the yo-yo, made of wood in the 1960s. $30-40

Duncan Shrieking Sonic Satellite, blue, slanted writing "Duncan Satellite," made of wood in the 1960s. $25-35

Duncan Shrieking Sonic Satellite, on card, white with pink rims, sold originally for $1.00, made of wood in the 1960s. $60-70

Duncan Imperial on a large card, sold originally for $1.00, amber, made of plastic in the 1960s. $25-35

Duncan Imperial on a small card, No. 400, sold originally for $1.00, made of plastic in the 1960s. $25-35

Duncan Imperial, marbleized plastic with a fleur-de-lis logo, made in the 1960s. $35-45

Duncan Imperial Jr. Beginners on a card, sold originally for 49 cents. This yo-yo features a picture of a cat, dog and the Duncan yo-yo man. It was made of plastic in the 1960s. $60-70

Duncan Mardi Gras on a small card, No. 700, sold originally for $1.00, red, white and blue, made of plastic in the 1960s. $85-95

Duncan Mardi Gras on a large card, blue gold, sold originally for $1.00, made of plastic in the 1960s. $85-95

Duncan Little Ace Mobil advertisement. This yo-yo has a different seal than other Duncan Little Aces. It was made of plastic in the 1960s. $25-35

Duncan Little Ace, the name "Little Ace" appears in curved writing. This yo-yo was made of plastic in the 1960s. $25-30

Duncan Fun Pack: mixed trick book, yo-yo, wood top, and handball. The items feature pictures of the Campbell's Soup kid and were made in the 1960s. $70-80

DUNCAN

Duncan basketball on card, sold originally for $1.00, part of a sports line made of plastic in the 1960s. $20-35

Duncan baseball on card, sold originally for $1.00, part of a sports line made of plastic in the 1960s. $20-35

Above: Duncan Disney World of Color yo-yo on card, butterfly shape, sold originally for $1.00, made of embedded plastic in the 1960s. $40-70

Left: Duncan Bo-Yo bowling ball sports line, reads Official Bo-Yo by Duncan on one side and Amflite Bowling Balls on other side. This sports line yo-yo was made of plastic in the 1960s. $20-30

Right: Duncan Mickey Mouse Club little ace yo-yo, made of plastic in the 1960s. $30-50

Duncan World of Color concave sides yo-yo with stars embedded in the plastic, made in the 1960s. $35-45

Duncan Walt Disney Character Return Top yo-yo, Duncan Jr. Imperial model decorated with a 3-face Mickey design, sold originally for 49 cents. This yo-yo was made of plastic in the 1960s. $30-50

Duncan Mickey Mouse Club yo-yo, Duncan Jr. Imperial model adorned with a Mickey and Minnie design, sold originally for 59 cents, blister card. This Jr. Imperial was made of plastic in the 1960s. $30-50

Duncan Butterfly, early two-tone, large butterfly with gold lettering in the middle, made of plastic in the 1970s. $15-20

Duncan Butterfly, two-tone, small butterfly with gold lettering outside, made of plastic in the early 1970s. $15-20

DUNCAN

This Duncan Junior Yo-Yo Return Top came in Post cereal boxes. It is a two-tone teeny with concentric circles on one side. It was made of plastic in the 1960s. $10-15

Duncan Space Seattlite Needle made of wood in the 1960s. $150-175

This Duncan Junior Yo-Yo Return Top came in Post cereal boxes. It is a teeny with two-tone decoration and a label on both sides. It was made of plastic in the 1960s. $10-15

Duncan Space Seattlite Needle back side.

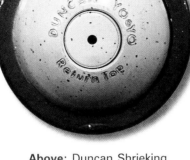

Above: Duncan Shrieking Sonic Satellite, two-tone white and blue, made of wood in the 1960s. $30-40

Right: Duncan Shrieking Sonic Satellite, red, in package, made of wood in the 1960s. $65-75

Duncan Satellite Mark 1, flat sides, red, made of wood in the 1960s. $30-40

Duncan Shrieking Sonic Satellite, blue, in package, made of wood in the 1960s. $65-75

Duncan Tournament cross-flags, red, in package, made of wood in the 1960s. $30-45

Duncan Tournament cross-flags, black, in package, made of wood in the 1960s. $30-45

Duncan Tournament cross-flags, pink, in package, made of wood in the 1960s. $30-45

Duncan Tournament cross-flags, yellow, in package, made of wood in the 1960s. $30-45

Duncan Tournament cross-flags, light blue, in package, made of wood in the 1960s. $30-45

Duncan Personalized Butterfly pack, pink, in package, trick book, two strings, gold letters, made of wood in the 1960s. $90-100

Duncan Personalized Butterfly pack, orange, in package, trick book, two strings, gold letters, made of wood in the 1960s. $90-100

Duncan Personalized Butterfly pack, green, in package, trick book, two strings, gold letters, made of wood in the 1960s. $90-100

Duncan Mardi Gras, red, white and blue, made of plastic in the 1960s. $65-75

Duncan Mardi Gras, in package, red, made of plastic in the 1960s. $90- $100

Duncan Mardi Gras, pink and silver, made of plastic in the 1960s. $65-75

Duncan Mardi Gras, green shades, made of plastic in the 1960s. $65-75

Duncan Super Tournament, light blue, made of wood in the 1950s. $30-40

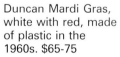

Duncan Mardi Gras, white with red, made of plastic in the 1960s. $65-75

Left: Duncan Super Tournament, white with green stripe, made of wood in the 1950s. $30-40

Center: Duncan Super Tournament, neon orange, made of wood in the 1950s. $30-40

Right: Duncan Super Tournament, orange with black stripe, made of wood in the 1950s. $30-40

Left: Duncan Super Tournament, yellow with silver stripe, made of wood in the 1950s. $30-40

Center: Duncan Super Tournament, natural with red stripe, made of wood in the 1950s. $30-40

Right: Duncan Super Tournament, navy with white stripe, made of wood in the 1950s. $30-40

Duncan Super Tournament, coral with silver stripe, made of wood in the 1950s. $30-40

Duncan Super Tournament, pink with black stripe, made of wood in the 1950s. $30-40

Duncan Super Tournament Return Top, pink with red stripe, made of wood in the 1950s. $35-45

Duncan Little G Genuine Tournament, lavender with pink stripe, made of wood in the 1940s-1950s. $40-50

Duncan Little G Genuine Tournament, pink with gold stripe, made of wood in the 1940s-1950s. $40-50

Duncan Super Tournament Return Top, red with black stripe, made of wood in the 1950s. $35-45

Genuine Duncan No. 77 Yo-Yo Tournament Top, natural wood, made of wood in the 1950s. $65-75

Genuine Duncan No. 77 Yo-Yo Tournament Top, day-glo yellow, made of wood in the 1950s. $70-80

Duncan Imperial on card, translucent blue, made of plastic in the 1960s. $25-35

Above: Duncan, shows Duncan yo-yo man's head, red and blue, made of wood in the 1960s. $15-20

Left: Duncan Imperial on card, matte black, made of plastic in the 1960s. $25-35

Right: Duncan, advertising Eversweet Orange Juice, Duncan yo-yo man's head, marked beginner's model on back, made of wood in the 1960s. $25-35

Festival

Festival Professional Model Basketball, part of the "Be-a-Sport" series. It has a festival logo on one side and is labeled "Official League" on the other. It was made of plastic in the 1960s-1970s. $20-30

Festival Professional Model Baseball, part of the "Be-a-Sport" series. It has a festival logo above the baseball "seams" and is labeled "Official League." This yo-yo was made of plastic in the 1960s-1970s. $20-30

Festival Professional Model Football, part of the "Be-a-Sport" series. This yo-yo has a festival logo on one side and is marked "Official League" on the other. It was made of plastic in the 1960s-1970s. $20-30

FESTIVAL

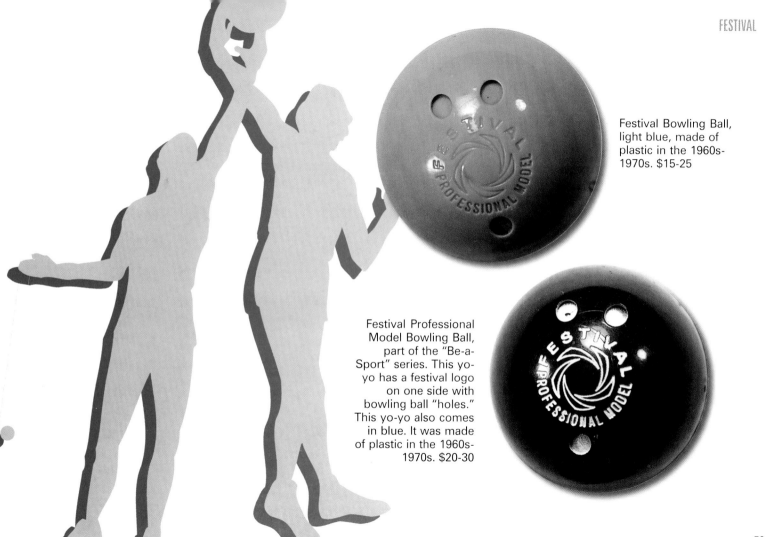

Festival Bowling Ball, light blue, made of plastic in the 1960s-1970s. $15-25

Festival Professional Model Bowling Ball, part of the "Be-a-Sport" series. This yo-yo has a festival logo on one side with bowling ball "holes." This yo-yo also comes in blue. It was made of plastic in the 1960s-1970s. $20-30

FESTIVAL

Festival Professional Model Golf Ball, part of the "Be-a-Sport" series. It has a festival sticker, is sealed in the box, and was made of plastic in the 1960s-1970s. $20-30

Festival Professional Model 8 Ball, part of the "Be-a-Sport" series. This yo-yo has an "8" in circles in the plastic on one side and a festival logo on other. This example is in the box and was made of plastic in the 1960s-1970s. $15-25

Fli-Back

Fli-Back yo-yo, two-tone red and blue with a white circular paint seal, made of wood in the 1970s. $10-15

Fli-Back 45, gold die stamped 45 with stars circling, no brand name but this example was made by Fli-Back. It was made of wood in the 1960s. $35-45

Fli-Back yo-yo, undersized red with circular foil sticker, made of wood in the 1970s. $20-30

Fli-Back yo-yo, blue with circular foil sticker—same the red example, made of wood in the 1970s. $20-30

Fli-Back yo-yo, blue translucent plastic with a white circular paint seal, made of plastic in the 1960s-1970s. $15-20

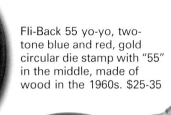

Fli-Back 55 yo-yo, two-tone blue and red, gold circular die stamp with "55" in the middle, made of wood in the 1960s. $25-35

Fli-Back 60 yo-yo, two-tone blue and red with a gold circular paint seal with "60" in the middle, made of wood in the 1960s. $35-40

Fli-Back 65 yo-yo, two-tone red and blue with a gold circular die stamp with "65" in middle, made of wood in the 1960s. $35-40

Fli-Back yo-yo, red and blue undersized model with straight lettering in a circle, made of wood in the 1960s-1970s. $10-15

Fli-Back Champion Style 55, gold die stamp, made of wood in the 1960s. $15-20

Fli-Back Top, eagle on die stamp, made of wood in the 1960s. $15-20

Fli-Back box of Championship Tournament Yo-Yos, 3 colors in plastic translucent sparkle with a circular white paint seal, made by Sock-It Co. These yo-yos were made in the 1960s. $75-100

Fli-Back Return Top Genuine Tournament, foil seal with eagle logo, blue, made of wood in the 1950s. $45-55

Fli-Back Return Top Genuine Tournament, foil seal with eagle logo, green, made of wood in the 1950s. $45-55

Flores

Flores yo-yo made of wood, c. 1928. $350-450

Forrester Worlds Fair yo-yo: undersized and regular size models were made. This example has a gold paint stamp and was made of wood in the 1960s. $15-25

Forrester

German Goody

German rattler, white and blue with red on the other side, made of tin in the 1930s. $40-50

Goody Winner, green with paint stamp and 1 jewel in the middle, made of wood in the 1950s-1960s. $65-100

Goody Master, blue with paint, stamp and 1 jewel in the middle, made of wood in the 1950s-1960s. $65-100

Goody Master, plain wood with paint stamp and 1 jewel in the middle, made in the 1950s-1960s. $70-110

Goody Filipino Twirler, purple, made of wood in the 1950s-1960s. $30-40

Goody Filipino Twirler, black with a different stamp, made of wood in the 1950s-1960s. $30-40

Goody Filipino Twirler junior-sized model with a horse on the back, made of wood in the 1950s-1960s. $35

Top left: Goody Filipino Twirler in a different size with a different stamp, made of wood in the 1950s-1960s. $35-45

Top right: Goody Filipino Twirler, light blue with red on the other side, made of wood in the 1950s-1960s. $30-40

Bottom left: Goody Filipino Twirler, yellow and purple and a paint stamp, made of wood in the 1950s-1960s. $35-45

Bottom right: Goody Filipino Twirler, many variations of the paint stamp design exist, two-tone yo-yo made of wood in the 1950s-1960s. $30-40

Gropper

Gropper Up-N-Down Top, gold die stamp, made of wood in the 1950s. $35-45

Gropper Up-N-Down Top, gold stamp, made of wood in the 1950s. $35-45

Gropper Up-N-Down Top, single jewel on back, made of wood in the 1950s. The authenticity of this model is questionable. $40-50

Hasbro

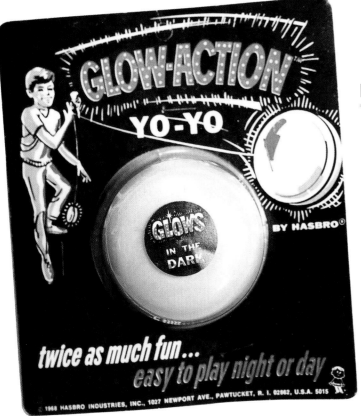

Hasbro Glow-Action Yo-Yo, glow plastic on a card with a sticker on the outside of the card. This model was made of plastic in 1968. $20-30

Hasbro pencil with a yo-yo, made in the 1960s. $10-15

Hi-Ker

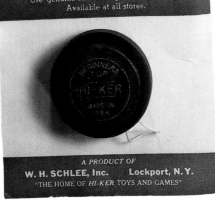

Hi-Ker Beginners Top on card, die stamp, undersized, two-tone, made of wood in the 1950s. $35-50

Hi-Ker Professional, lime green, made of wood in the 1950s. $55-65

Hi-Ker Professional Top, gold die stamp, airbrush stripe, made of wood in the 1950s. $55-65

This Hi-Ker Spin Master has glitter paint and a bow-tie shaped gold die stamp. Hi-Ker also made a Spin Master with a different seal. This model was made of wood in the 1950s. $55-65

Hi-Ker Flat Top butterfly design, gold flake paint, gold die stamp, made of wood in the 1950s. $60-70

Left: Hi-Ker Flattop, light green, made of wood in the 1950s. $60-70

Center: Hi-Ker Flattop, red, made of wood in the 1950s. $60-70

Right: Hi-Ker Flattop, orange, made of wood in the 1950s. $60-7

Hy-Lo yo-yo made of cast iron in the 1920s. $100-150

Hy-Lo

Jack Russell

Jack Russell's Super Yo-Yo with Coke advertising and convex sides, made of plastic in the 1970s. $15-20

Ka-Yo

Ka-Yo Whistling wood grain yo-yo with litho and paper seal, made of tin in the 1930s-1940s. $180-230

Ka-Yo Musical tin whistler yo-yo decorated with a litho and made in the 1930s-1940s. $175-225

Kaysons Novelty Co.

Kaysons Novelty Co. Streamline Top, paper decal, made of wood in the 1930s. $40-50

Kaysons Novelty Co. Streamline Top, paper decal, blue/orange, made of wood in the 1930s. $40-50

Knights

Knights yo-yo, gold decal seal, made of wood in the 1950s. $30-40

Kusan

Kusan Twin Twirler, green swirled plastic. This is the Falcon model, a smaller version of the Twin Twirler with a ribbed top part. It was made of plastic in the early 1960s. $45-55

Kusan Twin Twirler, a larger smooth version called the Bat model, blue translucent plastic. The third model is known as the Flying Eagle. This Twin Twirler was made of plastic in the early 1960s. $45-55

L.H. Knibb Co.

L.H. Knibb Co. Dyna-Glow Skil Top, segmented glow decoration, made of plastic in the 1950s. $60-70

Lumar

Lumar #33 Junior Express, manufactured in Britain by Louis Marx. The Junior Express was made of tin. Some examples are as old as the 1930s. $80-90

Marx

Marx, Magic Marxie Majestic Yo-Yo, translucent with gold stamp, mounted on card, made of plastic in the 1960s. $10-20

Medalist

Genuine Medalist Trickmaster Yo-Yo (on Medalist Trophy card) with gold stamp, made of wood in the late 1960s. $50-60

Miscellaneous and unknown

Sparkling Glo-Yo with a flint on the inside of the yo-yo to make it spark. Pat No. 1949858. This yo-yo was patented by Irving C. Brown and made of tin in 1934. $75-100

The Lord Is My Shepherd yellow paint seal adorned yo-yo made of wood in the 1950s. $25-35

MISCELLANEOUS

No name blue and yellow 1 1/2" wooden Japanese yo-yo, c. 1950s-1960s. $3-6

No name red and white plastic 2" yo-yo. $2-3

Red Top 2 1/8" steel yo-yo made in Japan in the 1950s. $20-30

MISCELLANEOUS

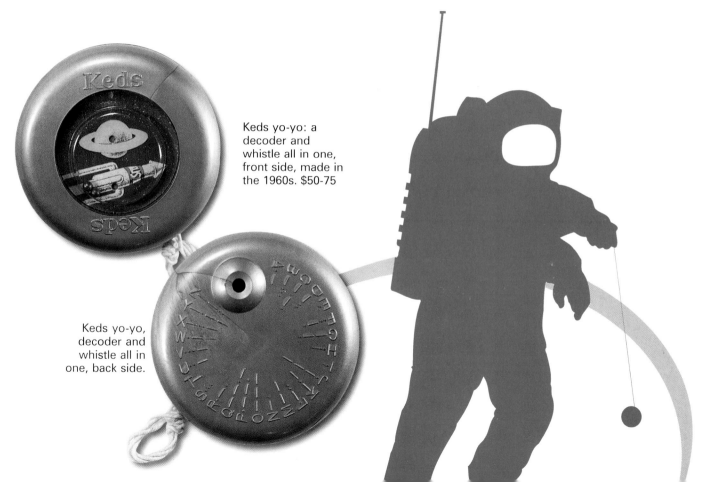

Keds yo-yo: a decoder and whistle all in one, front side, made in the 1960s. $50-75

Keds yo-yo, decoder and whistle all in one, back side.

MISCELLANEOUS

"Original 1955 Yo-Yo," a new Duncan yo-yo made in 1995. $10-15

MISCELLANEOUS

Wooden no name gum ball prize yo-yo, yellow and blue, made in Japan in the 1950s. $3-5

No name wooden gum ball prize yo-yo, red and yellow, made in Japan in the 1950s. $3-5

No name wooden yo-yo, red. $2-3

Duncan Yo-Yo winner patch, shield shape, 5 x 5 3/4 inches, made in 1940. $100-125

101

MISCELLANEOUS

Second Duncan Yo-Yo Contest patch, shield shape, red and white, made in the 1960s. $25-35

Duncan District Champion patch, circle shape, made in the 1950s. $100-125

Duncan Junior Instructor 20 yo-yo trick patch made in the 1950s. $50-75

MISCELLANEOUS

Duncan Yo-Yo Award pennant made in the 1950s. $25-35

Above: Cheerio 24-trick award patch, shield shape, Canadian model, made in the 1950s. $100-125

Right: Cheerio winner award patch, maple leaf shape, Canadian model, made in the 1950s. $150-200

MISCELLANEOUS

Above: Royal Yo-Yo Champion eagle patch made in the 1950s-1960s. $40-50

Right: Duncan trick book, 5 x 7 inches, made in 1961. $15-25

Left: Duncan Extra Tournament Tricks, originally 10 cents, made in 1960. $15-25

Below: Duncan How To Master Championship Tricks, made in 1947. $35-45

MISCELLANEOUS

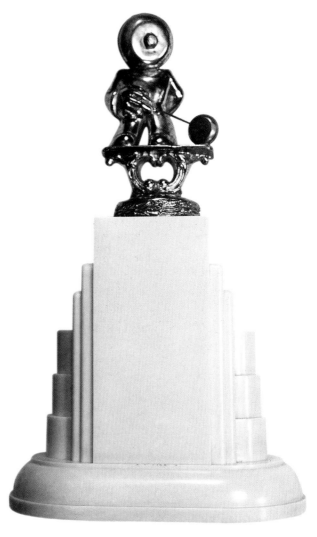

Above: Duncan The Art of Playing Yo-Yos, made in 1950. $35-45

Right: Duncan Yo-Yo trophy won by Mike Taylor: missing gold plate, made in the 1950s. $250-300

N.N. Hill Brass Co.

N.N. Hill Brass Co. Rattling yo-yo with a paper sticker, made of tin in the 1940s. $85-95

Nadson

Genuine Nadson Twirler Top, jeweled (jewels may be different colors): a wooden yo-yo made in Hong Kong in the late 1950s-early 1960s. $55-65

Novel Swing Top yo-yo, gold die stamp, made of wood in the 1930s. $40-50

Novel

Parker

Parker Original Genuine Pro Tournament Spinner Yo-Yo Top marked with a gold die stamp. A Canadian company owned the Canadian copyright on this wooden yo-yo made in the 1980s. $10-15

Royal

Royal Official Tournament Yo-Yo Tops, gold die stamp with a chevron and crown logo, made of wood in the 1950s-1960s. $30-40

Royal Special Yo-Yo Tops, gold die stamp with a chevron and crown logo, made of wood in the 1960s. $25-35

Royal Champion Standard Top, two-tone, gold die stamp with a crown logo, made of wood in the 1940s-1950s. $30-40

Royal Champion Junior Top, full sized, two-tone, gold die stamp with a crown logo, made of wood in the 1940s-1950s. $30-40

Royal Yo-Yo Tops undersized two-tone with crown, chevron logo; gold die stamp made of wood in the 1950s-1960s. $20-30

Royal TV Picture Yo-Yo: a plastic yo-yo with clear, convex snap-on sides and lenses in glassine bags. This model was made of plastic in the 1960s-1970s. $25-35

Star

Star Return Top wooden yo-yo marked with an ink stamp and made in the 1960s. $20-30

Strings for Yo-Yos

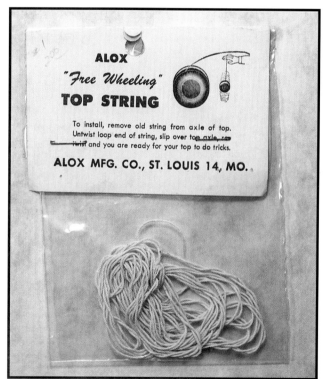

Alox Free Wheeling Top String, made in the 1950s. $20-30

STRINGS

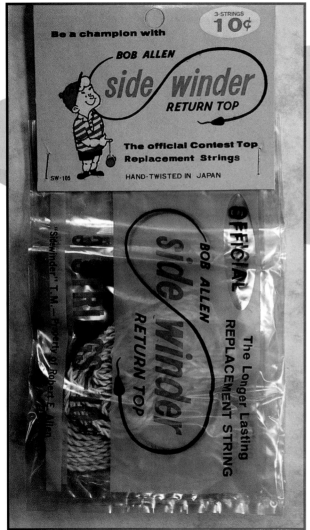

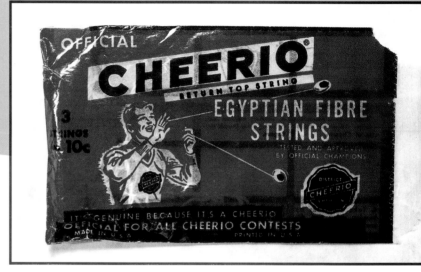

Above: A Cheerio string pack that sold originally for 3 for 10 cents. This pack was made in the 1950s. $25-30

Left: This Bob Allen string pack sold originally for 3 for 10 cents and was made in the 1960s. $15-20 each

STRINGS

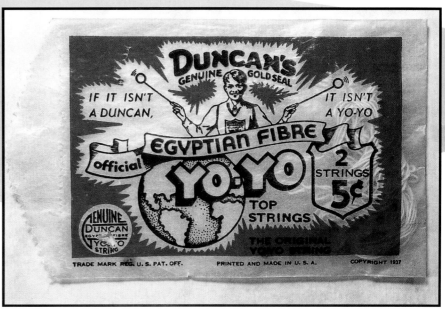

Above: Duncan string pack, small globe, sold originally for 2 for 5 cents, made from the 1930s-1950s. $25-35

Left: Chico Superb Chico Strings No. 220 sold originally for 2 for 5 cents and was made in the 1950s. $25-30

113

Above: Duncan Super-Cord string pack, yellow with an eagle, sold originally for 3 for 10 cents, made in the 1960s. $15-20

Right: Duncan string pack, vertical red coloring with the yo-yo man decoration, sold originally for 2 for 10 cents, made in the 1960s. $10-15

Duncan string pack, light blue with yo-yo man, sold originally for 2 for 10 cents, made in the 1960s. $10-15

Duncan string pack, red with yo-yo man, sold originally for 2 for 10 cents, made in the 1960s. $10-15

STRINGS

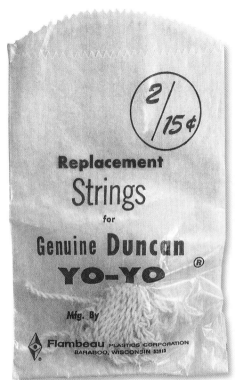

Right: Duncan string pack, white with red writing, vertical package with a picture of a yo-yo as decoration, sold originally for 2 for 15 cents, made in the 1960s. $10-15

Far right: Duncan string pack, white with red, vertical, no pictures, sold originally for 2 for 15 cents, made in the 1960s. $10-15

STRINGS

Duncan string pack card, red with yo-yo man, sold originally for 2 for 10 cents each, 6 to a card, made in the 1960s. $60-80

Festival String pack made in the late 1960s. $10-15

117

STRINGS

This Fli-Back String pack sold originally for 3 for 10 cents when it was made in the 1960s. $10-15

The packaged Hi-Ker Strings No. 5 sold originally for 2 for 5 cents when manufactured in the 1940s. $35-45

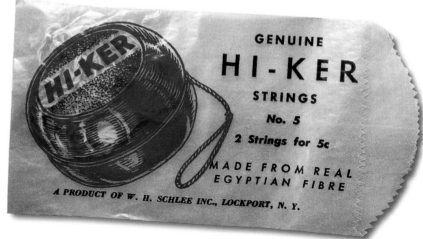

STRINGS

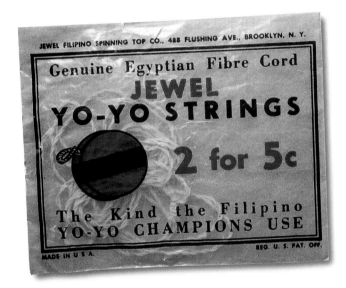

Jewel Yo-Yo strings sold 2 for 5 cents when made in the 1950s. $30-35

Joe Radovan string pack No. 207 sold originally for 2 for 5 cents made in the 1950s. $20-25

STRINGS

Above: Kaysons Streamline Top Strings sold originally for 3 for 5 cents when made in the 1930s. $30-40

Right: Monarch string pack No. 215 sold originally for 2 for 15 cents when produced in the 1970s. $5-10

National string pack with 2 strings, made in the 1960s. $10-15

Royal Two Famous Filipino Yo-Yo Champions string pack No. 207, sold originally for 2 for 5 cents, made in the 1950s. $20-25

STRINGS

Left: Royal Yo-Yo Trick Cord No. 210, jeweled model displayed as package art, sold originally for 2 for 10 cents, made in the 1960s. $20-25

Right: Royal string pack No. 310, sold originally for 2 for 10 cents, chevron model displayed as package art, made in the 1970s. $10-15

Tico

This Tico Super Duper Return-O-Top giant yo-yo, on the original card, was made of plastic in the 1960s. $15-20

A Whirl King Top Standard Model marked with a large crown and made of wood in the 1960s. $15-25

Whirl King

A Whirl King Standard Model Return Top marked with a small crown and made of wood in the 1940s-1960s. $15-25

New York World's Fair

New York World's Fair yo-yo decorated with a 1939 World's Fair paper decal, purple & orange, one of several styles of yo-yos produced for the 1939 New York World's Fair, made of wood in 1939. $50-60

New York World's Fair yo-yo sporting a 1939 World's Fair paper decal, orange & black, made of wood in 1939. $50-60